Teaching Science as Inquiry

by
Steven J. Rakow

Library of Congress Catalog Card Number 86-61751
ISBN 0-87367-246-1
Copyright © 1986 by the Phi Delta Kappa Educational Foundation
Bloomington, Indiana

Table of Contents

Introduction 7

A Tale of Two Teachers 8

What Is Inquiry? 13
   The Nature of Science as a Process of Inquiry 14
   Teaching Science as a Process of Inquiry 15
   Learning Science as a Process of Inquiry 18

The Learning Cycle: A Model of Inquiry Teaching/Learning 21

The Status of the Inquiry Approach in Science Education 27
   Recommendations for Promoting the Inquiry Approach 30

References 32

Introduction

With the launch of the Soviet satellite, Sputnik, in 1957, great concern was voiced about the deplorable state of science and mathematics education in the United States. In response to these concerns, the federal government appropriated large sums of money to upgrade the teaching of science and mathematics, resulting in the development of several innovative curriculum projects. A common element in all of the new science curriculum projects was an approach known as inquiry, that is, the gathering of information by means of hands-on manipulation of materials.

This fastback will examine the use of the inquiry approach in the teaching of elementary science. Specifically, three dimensions of inquiry are considered: the nature of science as a process of inquiry, the teaching of science as a process of inquiry, and the learning of science as a process of inquiry.

My bias is that an inquiry approach is the best way to teach and learn science — a bias supported by several years of teaching science to students and teachers. Although certainly not the easiest way to teach science, anyone who has seen the excitement in children's eyes when they learn through this approach would find it difficult to teach in any other way. In this fastback, I shall strive to convince the reader of the value of the inquiry approach in the elementary science curriculum.

A Tale of Two Teachers

Elizabeth McDonald and Doris Smith are both third-grade teachers in a medium-sized, suburban elementary school. They have been teaching next door to each other for eight years. Both are committed teachers; however, each has a very different style of teaching.

Mrs. McDonald has loved science since she was very young. She communicates this love to her students by her enthusiasm when teaching science. Her room is an active place to learn. A visitor walking into her room sees no nice, neat rows of student desks; rather, the desks face each other in groups of four. The walls are covered with brightly colored posters on such topics as the Young Astronauts Program. The aquariums are located along the back wall, and a cage with a gerbil sits near the window.

Today the students are actively involved in making observations. It takes a moment to determine what it is that they are observing since their heads are pressed closely together in a tight group.

"One, two, three, four, five. There are five crickets."

"No. I counted six. There is one hiding behind the stick."

"Six," says Johnny as he records the number on his worksheet.

"For the past few days we have been observing our terrarium," Mrs. McDonald says. "Can anyone tell me what our terrarium needs to grow?"

"Light."

"Water."

"Air."

"Very good," replies Mrs. McDonald. "Then we added some crickets to the terrarium. You have had a chance to watch the crickets for a couple of days. Today I have something new to add to your terrarium." Excited murmurs are heard as Mrs. McDonald reaches into a paper bag and removes a clear plastic bag containing a chameleon.

"What's that?"

"It looks like a lizard."

"We have those in our backyard."

"Does anyone know what I have in this bag?" asks Mrs. McDonald. Several hands are raised. "Brenda."

"It's a chameleon."

"That's right. What do you know about chameleons?"

"They can change colors."

"They eat insects."

"Those are all good ideas. I want each group to add a chameleon to your terrarium and observe the chameleon carefully." Mrs. McDonald moves to each group to distribute a plastic bag containing a chameleon.

"How do we get it out?" Mary asks her group members.

"Just reach in and grab it."

"Don't squeeze too hard. You'll hurt it!"

After some hesitancy, Peter puts the chameleon into the terrarium. At first the chameleon stands still moving its head from side to side to survey the area. Slowly it begins to move about the terrarium as the children watch spellbound. Suddenly the chameleon lurches forward and grabs one of the crickets in its jaws.

"Mrs. McDonald. Mrs. McDonald. The chameleon ate one of the crickets."

"Oh, yech!"

"Now there are only five crickets," says Johnny, erasing his previous answer and recording the new number.

After giving the students time to observe the chameleons and the crickets, Mrs. McDonald tells them to return the terrarium to the shelf and requests their attention.

"I would like to teach you two new words," she says, writing the words *predator* and *prey* on the chalkboard. "This word is 'predator.' Can you all say that with me?"

"Predator," the class echoes.

"And this word is 'prey.' Can you all say that with me?"

"Prey," they respond.

"A predator is an animal that eats another animal. Can you think of any examples of predators?"

"The chameleon."

"A lion."

"A hawk."

"Yes, those are all good examples of predators. A prey is an animal that is eaten by a predator. Can you think of any examples of prey?"

"A cricket."

"A mouse."

"A small fish."

"Yes, those are all good examples of prey. Today at recess, we are going to play a game where some of you will be field mice and some of you will be hawks. It is called the Predator-Prey Game. Doesn't that sound like fun?"

"Yeah," the children respond.

Next door Doris Smith also is conducting a lesson on predator-prey relationships in animals. Mrs. Smith's approach to the lesson differs dramatically from Mrs. McDonald's. Her style reflects, in part, her lack of confidence in teaching science. Mrs. Smith took only the minimum number of science courses necessary to graduate from college. She has painful memories of one class where she was required to dissect a frog. She does enjoy teaching, however, especially teaching reading.

Her classroom is arranged much more conventionally — five rows across and six seats deep. There are posters on the walls featuring Newbery Award books and travel posters of interesting places to visit. In the front of the classroom is a student duty chart: class monitor, pledge leader, etc.

"In our science reading today, we will learn two new words," begins Mrs. Smith as she writes the words *predator* and *prey* on the chalkboard. "This word is 'predator.' A predator is an animal that eats other animals. A hawk is a predator because it hunts field mice. The other words is 'prey.' A prey is an animal that is eaten by other animals. The field mouse is an example of prey. Please open your science textbooks to page 48. Who would like to begin reading? Adam."

"Many animals live together in the forest. Some of these animals hunt other animals for food. These hunters are called pred..., pred..."

"Pred-a-tors," says Mrs. Smith, pronouncing each syllable.

"Predators."

"Who would like to continue? Rebecca."

"The animals that are eaten for food are called prey. Predators and prey are part of the life cycle of all living things."

The children continue reading until they have finished the section on predators and prey.

"Please take out a pencil," says Mrs. Smith. "I would now like you to complete this worksheet on predators and prey."

Mrs. McDonald and Mrs. Smith represent two distinctly different ways to teach science in the elementary school. It is not uncommon to find teachers in the same school with such varying approaches to the teaching of science. Which is the best way to teach science in the elementary school? That is a question that has been debated for many years — one that I will attempt to answer in this fastback.

During the late 1970s, groups of science educators met to examine the status of science education in response to a number of critical national reports that had been released. One group, chaired by Harold Pratt of the Jefferson County (Colorado) Schools, summarized the teaching of science in the elementary schools as follows:

> The typical elementary science experience of most students is at best very limited. Most often science is taught at the end of the day, if there is time, by a teacher who has little interest, experience or training to teach science. Although some limited equipment is available, it usually remains unused. The lesson will probably come from a textbook selected by a committee of teachers at the school or from teacher-prepared worksheets. It will consist of reading and memorizing some science facts related to a concept too abstract to be well understood

by the student but selected because it is "in the book." (Pratt 1981, pp. 73-74)

With this discouraging view of science teaching, it is little wonder that by the age of nine, large numbers of elementary students have been "turned off" to science. There are, however, alternatives to this textbook-centered approach to science teaching. The inquiry approach advocated in this fastback involves students in active, hands-on manipulation of their environment. It not only is an effective teaching approach but results in the development of positive attitudes toward science.

What Is Inquiry?

Since the curriculum reforms of the early 1960s, the teaching of science has departed radically from earlier methods in elementary science programs. Prior to the 1960s, the primary focus of the elementary science curriculum was teaching facts, concepts, and laws. The textbook was the basic source of information. Laboratory investigations, when used, served primarily to confirm concepts that had been read about or discussed in class.

The new approach to science teaching made laboratory experiences central. Students investigated and inquired about their world. Their own observations served as the authoritative source of data. They discovered the concepts and principles of science in much the same way as the original discoveries were made. The textbook served as a reference.

The inquiry approach used in these post-Sputnik science curriculum projects reflected a new partnership between educators and scientists. The emphasis on firsthand investigation in learning science reflected scientists' belief that students should model the processes used by scientists in developing scientific concepts. This is not a new notion. Both Socrates and Aristotle advocated the inductive approach to learning concepts. Aristotle developed guidelines for the collection and analysis of data that remain the basis for scientific inquiry today. In more recent times, Rousseau and Pestalozzi stressed the importance of direct observation in learning.

Perhaps the most direct influence on current science teaching was John Dewey. His notion of discovery as a method of acquiring knowledge closely parallels the inquiry approach advocated by the curriculum reformers of the 1950s and 1960s. As Dewey wrote:

> No one expects the young to make original discoveries of just the same facts and principles as are embodied in the science of nature and man. But it is not unreasonable to expect that learning may take place under such conditions that from the standpoint of the learner there is a genuine discovery. (Dewey 1916, p. 354)

This discovery approach espoused by Dewey places less emphasis on what information is learned and greater emphasis on the logical thinking processes by which new knowledge is acquired. With the rapid rate at which knowledge is expanding, it has become less and less tenuous to prescribe which concepts should be transmitted to students in their 12 or more years of formal education. Advocates of the discovery or inquiry approach argue that it makes more sense to teach students the logical thinking skills to construct knowledge. With these skills, students are better prepared to acquire new knowledge.

The discovery approach of Dewey was rechristened the "inquiry approach" during the curriculum reforms of the 1950s and 1960s. While Dewey focused on all teaching and learning as an inquiry process, the curriculum reformers also were interested in promoting the inquiry approach as a method of learning the discipline of science. There is a distinction between science as inquiry and the teaching and learning of science as inquiry. J.J. Schwab speaks to this distinction:

> The phrase the teaching of science as enquiry is ambiguous. It means, first, a process of teaching and learning which is itself an enquiry, 'teaching as enquiry.' It means, second, instruction in which science is seen as a process of enquiry, 'science as enquiry.' (Schwab 1961, p. 65)

The Nature of Science as a Process of Inquiry

When science is presented as a collection of facts, students are left with the perception that everything that there is to be known about science is already known and can be found in their textbook. This is a gross misrepresentation of the nature of science. Science is constantly changing. As new research and advanced technology become available, our understanding of the world is altered. History is replete with examples of scientific discoveries that dramatically changed our perceptions of the world.

In a technological society such as ours, it is imperative that citizens be able to cope with the rapid changes resulting from science and technology. An understanding of the fluid nature of science helps students to realize that change is the rule rather than the exception.

In addition to understanding the dynamic nature of science, students need to develop those attitudes that characterize the enterprise of science. Some of these attitudes are 1) curiosity, 2) willingness to suspend judgment, 3) open-mindedness, and 4) skepticism. Curiosity is the inspiration for all science. Scientists are question-askers; any observation they make may be the catalyst for an investigation. Scientists develop hypotheses and draw inferences from direct observation of natural phenomena and withhold judgments until all the data have been accumulated. They must be open-minded enough to set aside previously held views in light of new information.

Given the dynamic nature of science, scientists must be willing to challenge accepted dogmas. They must exhibit a healthy skepticism toward any conclusion that is not supported by careful observation. Throughout history the great scientific discoveries have been made by mavericks who dared to challenge accepted ideas. By presenting students with historical role models of such scientists as Copernicus, Darwin, and Pasteur, they begin to see how the attitudes of scientific inquiry exhibited by these great scientists led to discoveries that challenged the accepted ideas of their times and eventually changed the course of history.

Teaching Science as a Process of Inquiry

Teaching science as a process of inquiry requires behaviors and attitudes that for many teachers are contrary to the ways in which they traditionally have taught and contrary to the ways in which they were taught as students. In fact, the difficulty of changing teacher behavior and attitudes has been a major impediment to the large-scale adoption of the inquiry approach to teaching science. What are these behaviors and attitudes that contribute to successful inquiry teaching?

1. *Successful inquiry teachers model scientific attitudes.* The attitudes of successful scientists are also characteristic of successful science teachers. These teachers possess the attitudes of curiosity, willingness to suspend judgment, open-mindedness, and skepticism; and they model these attitudes in their interactions with their students. They show a natural curiosity about their world but are not afraid to admit when they don't know the answer to a question. Rather, they use that lack of information as an opportunity to investigate with their students. They are willing to accept a variety of viewpoints without passing judgment, thus modeling attitudes for students that encourage them to explore alternative ways of approaching problems.

2. *Successful inquiry teachers are creative.* Because many elementary science textbooks are not inquiry oriented, teachers must be able to adapt existing materials to an inquiry style of interacting with students. The inquiry approach does not lend itself to a straightforward presentation of facts and concepts; teachers and students frequently are required to create new approaches for solving original problems.

3. *Successful inquiry teachers are flexible.* Students need to explore various ways for solving problems. Through this exploration, students are able to develop successful problem-solving strategies. Teachers need to be flexible if their students are to have opportunities to try out alternative strategies or to explore different points of view. Teachers, accustomed to telling students what to do, often find this to be one of the most difficult skills to acquire. Teachers also must be flexible in allocating time by extending their lesson if students' interests call for more time.

4. *Successful inquiry teachers use effective questioning strategies.* Research conducted by Mary Budd Rowe has demonstrated that the nature of classroom discourse has a profound influence on the effectiveness of inquiry teaching (Rowe 1978). Specifically, it has been shown that the use of open-ended questions, wait time, and neutral praise stimulate students to think divergently, to give more complete answers, and to participate in classroom discussion.

Closed questions such as "What kinds of fish are in your aquarium?" or "How many legs does a spider have?" limit response. These questions only elicit facts. While facts are useful in learning, too often classroom questioning never goes beyond the level of factual recall. Open questions encourage a diversity of responses. Questions such as "What do you observe in your aquarium?" or "Tell me about your experiment?" stimulate divergent thinking and elicit a variety of responses. When conducting inquiry discussions, teachers should increase their use of open-ended questions.

Wait time refers to periods of silence following the asking of questions. Wait Time I is the period of silence following a teacher's question before the teacher calls on a student, rephrases the question, or answers the question himself. In the typical classroom, Wait Time I is about one second long. Wait Time II is the period of silence following the first student's response until other students respond or the teacher goes on with the lesson. Wait Time II also is about one second long. Research has demonstrated that increasing wait time to three to five seconds has positive effects on the inquiry climate of the classroom. Specifically, students' responses are longer and more elaborated, there is more student-to-student interaction, and responses from shyer or slower students are more frequent.

Neutral praise refers to the nonverbal reinforcement a teacher gives to a student's response by listening carefully to the student, nodding, or perhaps writing the student's response on the chalkboard. In inquiry teaching, it appears that the use of neutral praise is more effective than repeated verbal praise after every student response. This statement may seem contrary to the instinct of teachers, which is to give students frequent praise, and is not meant to discourage teachers from praising students. But in the specific context of an inquiry discussion, constant verbal praise can have a stifling effect. Overuse of verbal praise gives students an expectation that their responses should be as good as each previously praised student response. Thus, students may be unwilling to share ideas, especially speculative ideas, if they fear their responses will be found incorrect, irrelevant, or inappropriate. With neutral praise, all ideas can be shared; then later, the teacher, with the help of the students, can evaluate the merit of each response.

5. *Successful inquiry teachers are concerned both with thinking skills and with science content.* An inquiry teacher focuses not only on the acquisition of the facts and principles of science but also on the thinking skills needed to solve problems. These thinking skills, sometimes referred to as "science process skills," model the investigative processes used by scientists. Acquiring these process skills, while taking time away from the teaching of science content, contributes to lifelong learning by giving students the skills to learn independently.

Learning Science as a Process of Inquiry

In the inquiry approach, students use hands-on investigation and experimentation to learn about their world. The skills that students need to carry out investigations and experiments should be a major focus of an inquiry science program. These skills are collectively referred to as the science process skills.

One of the 1960s curriculum reform programs, Science — A Process Approach (SAPA), focused specifically on developing science process skills. The creators of the program developed a progression of skills that they believe are essential for successful science investigations. These skills fall into two groups. The first group, Basic Process Skills, consisting of eight skills, are introduced in grades K-3. They include observing, classifying, using space/time relationships, using numbers, communicating, measuring, predicting, and inferring. The next group, Integrated Process Skills, consisting of six skills, are introduced in grades 4-6. This group of skills includes formulating hypotheses, controlling variables, experimenting, defining operationally, formulating models, and interpreting data. By acquiring these skills, students will be able to "do science." Following are brief descriptions of these 14 skills:

Observing. Observing refers to the use of the five senses to gather data about objects and events. Use of the five senses is important; students should be encouraged to use more than just their sense of sight when observing and gathering data.

Classifying. Classifying involves the grouping of objects or events according to similar characteristics. Identifying and grouping by patterns of similarity is a frequently used skill in science.

Using Space/Time Relationships. Young children find it difficult to comprehend phenomena that are outside of their immediate temporal or spatial experience. By helping students develop an awareness of phenomena and events outside of their immediate environment, they come to understand concepts in astronomy or earth science, for example.

Using Numbers. Quantification is the essense of science. The ability to describe the world numerically is basic to all scientific endeavor. Science activities provide students with practical applications of the concepts they have learned in mathematics. This process skill serves as a link between science and mathematics.

Communicating. Communicating involves the use of spoken and written words, graphs, drawings, and diagrams to share information and ideas with others. Scientific discoveries have little value unless they are communicated to others. This process skill serves as a link between science and language arts.

Measuring. Scientists gather and share data about the world by using common standards of measurement: length (inches, feet, meters, light years); weight (pounds, grams); volume (quarts, gallons, liters); and time (seconds, hours, year). Students must be proficient in using these standards in order to communicate their results to others.

Predicting. Predicting for the scientist is forecasting future events based on observations and inferences. Accurate predicting requires that the scientist know many pieces of information and how they interact. For example, scientists are seeking more information about the nature of earthquakes so that they can predict future earthquakes before they occur.

Inferring. An inference is a logical thought process to show a relationship between two or more observations. Science seeks to identify relationships between phenomena by making observations and, on the basis of those observations, generating inferences about those phenomena.

Formulating Hypotheses. An hypothesis is an educated guess that is then tested experimentally. The formulation of hypotheses is a key skill in the scientific method.

Controlling Variables. In a science experiment, two conditions, alike in every way but one, are compared to determine the influence of the one missing element. In designing and conducting science experiments, students must be able to identify and control the variables in order to determine their effect on the experiment.

Experimenting. Experimenting combines all of the process skills used in conducting a scientific investigation.

Defining Operationally. An operational definition is a definition framed in terms of students' experiences. For example, defining an acid as any substance that turns blue litmus red is an operational definition.

Formulating Models. A model is a verbal, structural, or graphic representation of the physical world. Scientists develop models as a way of describing the world, then test and refine those models as more information becomes available.

Interpreting Data. Interpreting involves the analysis and synthesis of data to support or refute a hypothesis.

Central to all three dimensions of inquiry — science as inquiry, teaching science as a process of inquiry, and learning science as a process of inquiry — is the notion of discovery. The nature of science itself is a process of discovery. Teaching science as a process of inquiry requires teachers to set up learning environments in which students can engage in discovery. Finally, learning as a process of inquiry involves students in using science process skills to investigate and discover patterns in the world.

The Learning Cycle:
A Model of Inquiry Teaching/Learning

An inquiry model for the teaching and learning of science is the Science Curriculum Improvement Study (SCIS), one of the curriculum projects sponsored by the National Science Foundation in the 1960s. The project is based on Piaget's developmental stages with content and instructional sequence following a progression from concrete to abstract experiences. Lessons in the SCIS program follow a three-stage sequence referred to as the Learning Cycle. These three stages are exploration, concept introduction, and concept application (Atkin and Karplus 1962).

Exploration. In the exploration stage (sometimes referred to as "messing around"), students are given unfamiliar materials (for example, a set of geoblocks or a set of batteries, bulbs, and wires) and are instructed to find out everything that they can about the materials. This unstructured exploration leads students to ask further questions about the materials, which they may wish to investigate. Given the concrete operational orientation of most elementary students, this hands-on manipulation is an essential part of the learning process. When students first have been allowed to "follow their own agenda" with the new materials, they are more likely to follow the teacher's instructions in later stages.

When initiating the exploration stage, the teacher must make a clear task statement that is open-ended enough to allow students to follow a variety of strategies and interests yet specific enough to give them some direction. The challenge for the teacher is to avoid being too specific and stifling creativity and being too general and facing a barrage of "What-are-we-supposed-to-do?" questions. Contrary to earlier approaches to teaching science in which experiments are used to confirm previously learned concepts, in the exploration stage laboratory investigation is used to introduce a concept.

Concept Introduction. Following the exploration stage, the concept is introduced by lecture, discussion, reading, films, or investigation. During this stage the teacher introduces the vocabulary relevant to the concept by using the strategy of "concept invention." Concept invention is the process of defining a concept in terms of the hands-on experiences students have had during the exploration stage. For example, if students discover that adding certain chemicals to a "mystery" blue liquid turned the liquid yellow, the teacher might invent the following concept: "Any substance that turns mystery blue liquid to yellow is an acid."

Concept Application. The final stage of the Learning Cycle is the application of the concept to new situations, especially those that are relevant to students' lives. Increasingly science educators have been concerned that students see the practical applications of science and technology in their lives. The concept application stage allows students to transfer their knowledge to new learning experiences.

The original Learning Cycle has undergone some changes in label and structure, but the structure presented above has been found to be adaptable to most classrooms. Specifically, the structure accommodates the use of a science textbook during the concept introduction stage and, at times, during the concept application stage (although most elementary textbooks do not give adequate coverage to the role of science and technology in modern society).

The Learning Cycle is an especially useful model for implementing the inquiry approach. Students are engaged in active, hands-on inquiry during the exploration stage. Through this exploration, they raise questions that serve as a stimulus for further investigation. As they undertake further investigation, new concepts are introduced and applications of concepts are illustrated.

Following are two sample inquiry activities using the Learning Cycle approach. The first, in the physical sciences, introduces the concept of acid. The second, a biology activity, demonstrates the effect of environmental factors on an organism.

Investigating a Mystery Liquid

Purpose: To investigate a mystery liquid (an acid/base indicator).

Materials: For each student, a plastic tumbler, drinking straw, and bromthymol blue indicator in water solution (the mystery liquid).

For each group of 4 or 5 students, containers of water, vinegar, baking soda, alcohol, cream of tartar, ammonia, salt, and Alka Seltzer.

Paper cups for mixing (paint mixing trays will work well)

Caution: Tell the students not to taste the mystery liquid or anything else used in the experiment.

Exploration Phase

1. Provide each student with a drinking straw and a plastic tumbler ¼ filled with bromthymol blue solution. Challenge them to find out everything that they can about the mystery liquid.
2. When the majority of students have discovered that blowing into the liquid with their straw will turn it yellow, ask them to put the materials aside and prepare for a discussion.
3. Ask: What did you discover about the mystery liquid? What else could we do to find out more about the mystery liquid?
4. Based on the students' responses, the teacher may wish to have them design experiments to carry out further investigations on their own or to proceed with the following investigation as a class project.

Concept Introduction

1. Provide students with containers of mystery liquid, mixing containers, water, baking soda, and vinegar.
2. Challenge the students to find out what happens when you add water, baking soda, and vinegar to the mystery liquid.

3. Ask: What did you find out? Could we use mystery liquid to classify those substances that were added? (Yes — some substances turn the mystery liquid yellow and some do not.)
4. Have the students put the materials away.
5. Invent the concept *acid*. "Anything that turns mystery liquid yellow is an acid."
6. Provide the students with alcohol, cream of tartar, ammonia, salt, and Alka Seltzer (half a tablet is sufficient).
7. Have the students predict what will happen when each substance is placed in the mystery liquid.
8. After the students have made their predictions, provide them with mystery liquid to test their predictions.
9. After the students have tested their predictions, ask: "Were you surprised by any of the results?"

Concept Application
1. Applications of acids in the students' lives are numerous. Following are some common acids that students might identify. Have them discuss how these acids function in their daily lives.

    Citric acid — citrus fruits
    Ascorbic acid — vitamin C
    Hydrochloric acid — muriatic acid
    Sulfuric acid — battery acid
    Salicylic acid — aspirin
    Tannic acid — tea and coffee

2. Discuss such environmental implications of acids as acid runoff from mines and acid rain.

How Environmental Factors Affect Animals in the Classroom

Purpose: To investigate the effect of temperature on the respiration of fish.

Materials: For each group, one goldfish or feeder fish, a plastic tumbler, magnifying glass, outline drawing of a fish, marking pens, thermometer, ice cubes, hot water, and stop watch or clock with a second hand.

Exploration Phase

1. Provide each group with a tumbler of water containing a goldfish and a magnifying glass.
2. Show the students the outline drawing of a fish below.

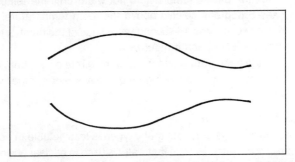

3. Instruct the students to make careful observations of the goldfish. As they observe features on the fish, add these to the drawing.
4. After the students have completed the drawing, use this to discuss the terms for the parts of the fish (mouth, eye, gills, fins, etc.).

Concept Introduction

1. Provide each student group with a tumbler of water containing a goldfish.
2. Ask the students to observe the relationship between the movement of the gills and the opening of the mouth. (They should observe the regular rhythm of mouth-gill-mouth-gill as the fish takes in water through its mouth and extrudes it past the gills where the dissolved oxygen is removed.)
3. Have the students observe and record the number of times that the fish opens its mouth in a 15-second interval. After they have had a chance to make several recordings, they should be ready to begin the experiment.
4. Record the temperature of the water at room temperature. Make two or three 15-second recordings of the number of times that the fish opens its mouth.

5. Place one or two ice cubes into the plastic tumbler with the fish until the temperature drops to 5 degrees celcius below the room temperature. Make two or three 15-second recordings of the number of times that the fish opens its mouth.
6. Place the plastic tumbler into hot water until the temperature rises 5 degrees celcius above the room temperature. Again, make two or three 15-second recordings of the number of times that the fish opens it mouth.
7. Plot the temperature and the breathing rate on a graph. Ask the students to predict the mouth-gill movement at other temperatures.

Concept Application

Discuss with the students that oxygen is less soluble in hot water than in cold water. Explain that water is used in nuclear power plants to cool the reactors. This water becomes hot and is then discharged into rivers. Ask students to predict what might happen to fish in such a river.

The Status of the Inquiry Approach in Science Education

Twenty-three prominent science educators met in 1978 under the auspices of Project Synthesis to examine the current status of science education (Harms and Yager 1981). This group examined several national reports and assessments; one subgroup focused specifically on the status of the inquiry approach in teaching science (Welch et al. 1981). Although the inquiry approach continues to be promoted by science educators, it appears that teachers increasingly are abandoning this approach and moving back to a textbook-centered approach to teaching science.

What do teachers perceive to be the limitations of teaching science as a process of inquiry? Following are the reasons the subgroup identified.

1. *Lack of training.* Following development of several new science programs in the 1960s, the National Science Foundation supported inservice training institutes for teachers. It appears that those institutes were effective in encouraging inquiry teaching; however, the NSF no longer supports such institutes. Elementary teachers; in particular, feel insecure about teaching science. Few elementary teachers have taken more than one or two college-level science courses; and it is not unusual for these future teachers to have had few, if any, college-level laboratory science courses. Given this lack of prepara-

tion, it is not surprising that many elementary teachers avoid teaching hands-on science.

2. *Lack of time.* Increasingly, teachers are under pressure to devote a major portion of their instructional time to teaching the basics of mathematics and reading, which leaves little time for hands-on science. It is unfortunate that "basics" are so narrowly defined. What could be more basic than the development of thinking skills, a prime objective of the inquiry approach in science?

3. *Lack of materials.* Inquiry science requires the manipulation of a great variety of materials by students. At the elementary level, many of these materials can be obtained from grocery, drug, and hardware stores; nevertheless, it still requires time for teachers to assemble and set up the materials. With the need to prepare lessons in several subject areas, elementary teachers find it difficult to find the time to plan for an active program of laboratory investigations. Some school districts have overcome this limitation by purchasing kits of laboratory materials assembled by science textbook publishers as supplements to their series.

The Minneapolis school system, under the leadership of science coordinator Joseph Premo, has developed an innovative approach for providing laboratory materials to elementary classrooms. It has assembled boxes of science materials, which are available for checkout by teachers, to accompany units in the adopted science program. Teachers simply send in their order and the laboratory materials are delivered to their school.

4. *Lack of support.* Along with a lack of training and lack of materials, many teachers feel a lack of support from their building or central office administration. Many principals who are not familiar with the inquiry approach to teaching science feel that if they have provided their teachers with a set of science textbooks, that should be adequate. Furthermore, only 20% of school districts in 1978 had a full-time science consultant (Weiss 1978). Without consultant support, teachers have nowhere to turn when faced with problems and often give up on the use of the inquiry approach.

5. *Over-emphasis on assessing content learning rather than process learning.* The type of evaluation used can have a powerful in-

fluence on both the content and method of learning. If teachers perceive that their students are to be evaluated on their ability to answer content-specific, multiple-choice items on a standardized science test, they might feel obligated to teach to the test. Admittedly, evaluating process skills is more difficult and requires more individualized means of assessment. Yet if teachers are not encouraged to consider inquiry skills as valued outcomes, they are unlikely to provide an inquiry climate for their students.

6. *Inquiry approach is too difficult.* Not only is an inquiry approach more difficult for teachers to manage, but it is perceived by many teachers as being too difficult for any but the brightest students. Actually, some of the early National Science Foundation programs were conceptually difficult for students; and they provided little guidance for teachers in helping students overcome these conceptual difficulties. Later versions of the programs have overcome several of these problems, but many teachers are unaware of these revisions.

The preceding discussion presents some of the obstacles to the use of inquiry in elementary school science. Nevertheless, creative teachers have been using the inquiry approach for a long time. The National Science Teachers Association established the Search for Excellence in Science Education to identify model science education programs (Penick 1983). One search focused on the inquiry approach in science and identified 10 exemplary programs. While these 10 programs represent only a sampling of all inquiry programs in operation, they do show some ways in which inquiry has been applied successfully in science classrooms.

Recently Texas has taken an important step in promoting inquiry teaching in science by legislating statewide minimum objectives ("essential elements") in all subjects and at all grade levels. Because Texas has statewide adoption of textbooks, these changes may also have a national impact (as goes Texas, so goes the nation). The legislated essential elements in science focus on process skills rather than content. Ten essential elements are common across the grade levels and science subjects. These essential elements focus on such process skills as observation, classification, measurement, and inference. Subdescriptors provide examples of content at specific grade levels or in specific science subject fields that would be appropriate for meeting the essential element. For example, the essential element that focuses on observation states: "The use of skills in acquiring data through the senses" (Huntsberger 1984). At the fourth-grade level, the subdescriptor for this essential element states: "The student shall be provided opportunities to observe phenomena and apply knowledge of facts and concepts and opportunities to observe that all living organisms depend on plants (food chains, food webs)." This focus on process, enforced by legislative mandate requiring that 40% of the junior and senior high science class time be laboratory oriented, is having a significant impact on science teaching in Texas.

Recommendations for Promoting the Inquiry Approach

In the final analysis, a renaissance in inquiry teaching in science will come when teachers believe that the outcomes justify the effort. The message from the Search for Excellence in Science Education exemplary programs is that the impetus for quality inquiry programs comes not from legislative or school district mandates but from creative and energetic teachers.

How does a teacher begin an inquiry program? Following are a few recommendations.

1. *Start small.* There is a temptation to want to revamp a science program completely to an inquiry approach. This is an overwhelming task and could be frustrating. A modest way to begin is to take one or two chapters of the science textbook and develop good inquiry activities for that chapter. Each year add a few more chapters. In a short time teachers will have a good collection of materials and activities.

2. *Enlist the aid of other teachers at the same grade level.* Colleagues who teach the same grade level can support and contribute to efforts to implement inquiry teaching. For example, if the task of collecting materials and developing inquiry activities for chapters in a science textbook is divided between two or more teachers, they can share their results; and soon all of the units of the textbook can be adapted for inquiry teaching.

3. *Look to the National Science Foundation curriculum projects for guidance.* Specifically, the Elementary Science Study and the Science Curriculum Improvement Study contain activities that are relatively easy to adapt to textbook programs. As a matter of fact, many currently used textbooks have been influenced by these NSF programs, so there is likely to be a good correspondence between the objectives of these programs and those of "second generation" texts.

Science opens a whole new world to elementary students. Given the opportunity to participate in hands-on activities in which they manipulate materials and observe natural phenomena, students learn the processes of science, which will serve them throughout their lives. Finally, the excitement that students feel toward science investigations and the awareness they gain about the role of science in their lives will help them to cope with the rapid change that science and technology impose on society.

Perhaps the best way to sum up the three dimensions of science inquiry — the nature of science as a process of inquiry, teaching science as a process of inquiry, and learning science as a process of inquiry — is simply to say, *science IS inquiry.*

References

Atkin, J.M., and Karplus, R. "Discovery or Invention?" *Science Teacher* 29 (1962): 45.

Dewey, John. *Democracy and Education.* New York: Macmillan, 1916.

Harms, Norris C., and Yager, Robert E. *What Research Says to the Science Teacher.* Vol. 3. Washington, D.C.: National Science Teachers Association, 1981.

Huntsberger, John P. "Essential Elements of Science Teaching." *The Texas Science Teacher* 13 (1984): 7-24.

Penick, John E., ed. *Focus on Excellence: Science as Inquiry.* Washington, D.C.: National Science Teachers Association, 1983.

Pratt, Harold. "Science Education in the Elementary School." In *What Research Says to the Science Teacher.* Vol. 3, edited by Norris C. Harms and Robert E. Yager. Washington, D.C.: National Science Teachers Association, 1981.

Rowe, Mary Budd. *Teaching Science as Continuous Inquiry.* 2nd ed. New York: McGraw-Hill, 1978.

Schwab, J.J. "The Teaching of Science as Inquiry." In *The Teaching of Science*, edited by J.J. Schwab and P.F. Brandwein. Cambridge, Mass.: Harvard University Press, 1962.

Weiss, Iris R. *Report of the 1977 Survey of Science, Mathematics, and Social Studies Education.* Research Triangle Park, N.C.: Center for Educational Research and Evaluation, 1978.

Welch, Wayne W.; Klopfer, Leopold E.; Aikenhead, Glen S.; and Robinson, James T. "The Role of Inquiry in Science Education: Analysis and Recommendations." *Science Education* 65 (1981): 33-50.